D1420668

AT BRECHIN WITH STIRKS

SOURCES IN LOCAL HISTORY No1

The Sources in Local History series is
sponsored by the *European Ethnological
Research Centre* c/o the National Museums
of Scotland, Queen Street, Edinburgh
EH2 1JD

General Editor: Alexander Fenton

AT BRECHIN WITH STIRKS

A Farm Cash Book
From Buskhead, Glenesk,
Angus, 1885-1898

Edited by
Alexander Fenton

CANONGATE ACADEMIC
in association with
The European Ethnological Research Centre
and the National Museums of Scotland

Sources in Local History No 1

First published in 1994 by
Canongate Academic, an imprint of Canongate Press Ltd,
14 Frederick Street, Edinburgh EH2 2HB
Copyright © The European Ethnological Research Centre

Financial support from the Greta Michie Bequest, *Scotland
Inheritance Fund*, is gratefully acknowledged.

British Library Cataloguing in Publication Data

A catalogue record for this book is available on request from
the British Library

ISBN 1 898410 09 7

Typeset by Hewer Text Composition Services, Edinburgh
Printed and bound in Great Britain by Bookcraft Ltd,
Midsomer Norton, Avon

Contents

Foreword

The story of the countryside and of farming in the later years of the nineteenth century is not as well known as it should be. There is a need for source material to be made available, with full texts for analysis.

In the 1880s–1890s, ways of working the land, and the lives of rural communities, were still much as they had been for decades before. This was especially true of a corner of Scotland like Glenesk, in a valley running into the Grampian Hills. It is not as remote as it seems, because it was for long a through route for travellers on foot or on horseback moving between Royal Deeside in Aberdeenshire and the Angus Lowlands. But the building of turnpike roads increasingly took people along the longer, though now quicker, east-coast route and Glenesk in the late nineteenth century was probably a quieter place than it had been a century before.

The Cash-Book is a record of the money outlay and income of a small farm which was working, as most did, at the lower margins of profitability, but also producing food, the value of which was not taken into the equation and is not easy to assess. Cash-books of this kind give only partial pictures of the economy. Yet their evidence is important. Economists, agricultural historians, ethnologists, rural sociologists, genealogists and others can make good use of such sources. General readers can browse through them, comparing these days with their own and getting a feeling for how a farmer of the 1880s–1890s lived and worked. Such sources also help to give some understanding of the view of the world that country folk had.

The European Ethnological Research Centre is planning a series of such sources, under the general title of 'Sources in Local History'. They are sometimes described as 'ego-documents', written by individuals for their own purposes, and not for the wide world to stare at. It is precisely because of their individual quality, without the bias or pressure that an audience would provide (other than the writer's own family), that they become valuable historical documents.

The present volume is dedicated to Greta Michie, Margaret Fairweather Michie (1905–1985), and to all Glenners, past and present. Greta's love of Glenesk led to the collection that now constitutes the Glenesk Museum. This includes archive material as well as objects, and the Buskhead Cash-Book came to light as part of her work in the Glen. It is a pleasure to continue that work through the publication of the volume. It stands also as a tribute to her memory.

Alexander Fenton

Analysis of the Cash Book

The farm of Buskhead in Glenesk, Angus, extends to about 329 acres, of which only about twenty-eight acres (8.5%) is arable. It is currently tenanted by L Brand.

In November 1983, James Stewart (Jimmy Buskie) gave up the holding after his family had been in it for 150 years. About their predecessors, there are only a few scattered pieces of information. The Parish Records note that in 1820, Alexander Mitchell and his wife Jean had a daughter born there and baptised Elizabeth. In 1857, Ann Milne, widow of Peter Copland in Shinfur, died at Buskhead, aged eighty-six. Earlier still, a David Low was in Buskhead in 1716. He paid in rent 50 merks of silver duty, 2½ merks for cess, 15/- for watch money, a merk for butter, 2 merks for poultry and 20/- for a part of a wether.

The 150 years of Stewart tenancy is a mark of the stability of occupation of Glen farms. From this Cash-Book, much can be gleaned about the workings of the farm. It was kept by Archibald Stewart from October 1885 until November 1898.

In 1841, the Census Returns list Archibald Stewart as being a farmer in Buskhead. He was thirty years old, as was his wife Margaret, and they had a son Alexander, aged one, and a daughter Jean Eggo, aged five. There were then two female servants, Mary Christison, aged fifteen, and the even younger ten-year-old Elisabeth More.

Five other individuals were listed under Buskhead: Joseph Gall (eighty), tailor, and his wife Christina (sixty-five); David Gall, shepherd (presumably their son), and James White, carrier. There were also two old ladies, Jannet Thow (seventy-five) and Agnes Milne (seventy), both listed as being of independent means.

By 1891, the range of people associated with Buskhead had become limited to three: Archibald Stewart (thirty-four), son of the Archibald listed in 1841, his widowed mother Margaret H., now seventy-nine, and a servant lass, Ann Low, aged eighteen. It is this Archibald who compiled the Cash Book, between the ages of twenty-nine and thirty-eight.

It is a hard-covered Cash-Book, complete except for one page covering expenditure from March to July 1897. It was reused in 1934, seemingly to record prices of groceries, and in 1936–40 to record sales of eggs. It is simply laid out, with income on the left-hand and expenditure on the right-hand pages. Items sold are relatively few; those purchased are more varied and take up more space.

Over the period covered (excluding the three months of 1885) there were fluctuations in income, 1886 and 1892 being poor years, but otherwise the pattern remained fairly steady. The average was £119. The worst figure was £54.6.6. (1886) and the best £152.7.9 (1898). Expenditure was on the whole more steady than income with an annual average of £117, a low of £94.13.8 (1887 – following a bad year) and a high of £150.5.4 (1898). The graphs of both income and expenditure were clearly rising as the end of the nineteenth century approached, but the increase in one was matched by the increase in the other. It cannot be said that the farm was demonstrably better off in 1898 than in 1886, and as the averages show, the difference between profit and loss over the thirteen year period stood at around £2. Thus bad weather, animal disease or other problems could have an unduly heavy effect. However, the Cash Book deals only with cash, and does not take into account the value of food from land and garden which, if it were calculated, would make the profit levels seem more comfortable.

Calculation of income and expenditure month-by-month reveals a rhythmic pattern. There is an expenditure peak in May and a sales peak in October, the latter reflecting in particular the successful outcome of the year's crop of sheep and wool. The major sources of income were from the sale of ewes, cast ewes, ewe hogs, wethers, lambs, gimmers and rams; wool and skins; cows, stirks, heifers, bullocks and stots; corn, by which is meant oats; grass seed; winter keep of a horse for Lord Hindlip; one sale of a foal; and one sale of 12 stones of pork at 5/- a stone.

Analysis shows that sheep provided most money, followed by cattle.

Sales as a Percentage of Total Income, 1886–98

Sheep	43.7%	Keep of Horse	4.5%
Cattle	34.5%	Grass Seed	1.0%
Wool	11.0%	Sale of foal	0.6%
Grain	4.5%	Sale of pork	0.2%

Sheep and wool, when added together, gave well over half the regular returns, marking the pastoral emphasis in the hills of the Glen. In terms of the narrow profit margin, extras like the keep of Lord Hindlip's horse were by no means insignificant.

The rhythm of the income pattern shows two high peaks; the rhythm can also be related to the nature of the goods sold and to the seasonal round of sowing, planting and cropping, the serving of cows and calving, and the wintering and summering of stock.

The farm had a number of fixed outlays. The rent of £15 a year, paid to the Earl of Dalhousie in equal instalments in May and November, remained constant throughout the period. Fire insurance, paid in February, rose from 15/- in 1886 to £1.4/- by 1898. Poor and school rates fluctuated between 6/5½ and 7/7 during 1886–96, then fell to 2/9½. Road money rose from 3/9 to 4/1 in 1889; in that year County Councils were established and such dues were then covered by the county rates, which started at 5/10 in 1890 and sank to 2/- in 1897. All parish and county dues were paid in December. To cover these outlays, the annual income had to show a basic profit of just under £17.

Wages also meant regular payments at the May and November terms, and also for casual work at other times of the year. The permanent staff that took up the bulk of the wage bill were women, presumably working mainly in the house. They were:

1886	Jane Christison	£11
1887–89	Rachel Burness	£10.10/- rising to £12
1890–94	Ann or Annie Low	£1 rising to £14.
	She left in August 1894, outside the regular term time.	
1894–95	Bella Hunter	£10
1896	Robina Middleton	£7.10/- for 6 months
	Jane Henderson	£7.10/- for 6 months
1897–98	Susan Grant	£8 for 6 months of each year
	Miss Tosh, housekeeper	£7.10/-

Wages rose from £11 to around £16 a year over the thirteen-year period. The appearance of a housekeeper towards the end was caused by a death in the family, noted in the Cash Book by a number of entries laconically marking illness and death.

> 9 Dec. 1892, Dr Ironside, £1.18/-; 6 Jan. 1896, Dr Ironside, £1.12/-; 4 Oct. 1898, Mrs Gall, nurse, £2; 13 Oct. 1898, Stewart Porter making grave, 8/-; 12 Nov. 1898, David Mitchell Joiner for Coffin, £3.5/-

Temporary or casual help was also got for a variety of purposes. Alex Middleton worked for part of each year from 1886 till 1890, at general duties. He was followed by John Valentine in 1890, and by Ewan McIntosh, also called 'Tosh', in 1891–92. His work included lambing ewes. Then came Archie Stewart, described as 'herd' and 'boy', in 1893–94; Griffeth Stewart in 1894–95; James Mearns in 1897; and James Gibson in 1898. On three occasions, 'Archd. Stewart' is entered. In June 1892 he received £16.10/- for thirty-three weeks of work, 15 October 1891 to 2 June 1892; in May 1896, £14 for twenty-eight weeks; and a further £13 in April 1898. In view of the irregular timing of these payments, in both years and months, it looks as if this is the writer of the Cash Book paying himself.

Harvest work was done by W Glass (who acted as mole-catcher at other periods), A Bowman, Jane Eggo, G Duncan, Mrs Duncan and others. The rate was up to 2/6 a day for men, and around 1/3 a day for women.

Turnip-hoeing occasionally needed an extra hand, such as A Birse who got 4/- for two days' work in July 1891. Potato-lifting made more regular demands, as on G Duncan in October 1891, and on J Stewart who got 4/- for two days in October 1895. Mrs Duncan was paid 4/- for four days' lifting turnips in November 1896.

Peat-working was the other great source of intensive work, mainly in June for cutting and raising, and to a lesser extent for building in July. The earlier job was harder, and D Ross, Andrew Christison, D Middleton and R Middleton got 2/- to 2/6 a day for doing it. The easier task of building was valued at only 1/9 a day, however, for G Duncan's two days.

There is one indication of shared herding of sheep, to judge by an entry on 26 May 1888: 'Paid Share of Shepperd's Wages herding the Ewes 18/6'.

Fixed outlays and wages took up 30% of the annual expenditure. The following gives the overall pattern.

Outlays as a Percentage of Total Expenditure, 1886–98

Sheep	24%
Wages	16%
Food and fodder (including milling, tradesmen and shops, oilcake and linseed meal, seeds and seed plants, and pigs)	15.23%
Fixed outlays (rent, dues, insurance)	24%
Horses and farm equipment (including blacksmith)	7.59%
Miscellaneous (mainly transport, veterinary, medical)	7.79%

4

Buildings and contents (including fuel, joiner and
ironmonger)   6.12%
Cattle   5.08%
Field maintenance (manure and lime, fencing and
dyking, mole-catching)   3.79%

Comparison of income and expenditure percentages shows that sheep
and wool led to the greatest concentration of annual effort; but the
income level of 54.7% has to be set against the 24% outlay, leaving a
30.7% input into the farm's economy. The proportions for cattle are
34.5% to 5.8%, leaving a 28.7% input. Even if the costs of fodder,
root crops, seeds and so on, are considered, the pattern is only
marginally affected, since they relate partly to sheep and partly –
though probably in slightly greater part – to cattle, as well as to pigs.
It is a matter of some interest that the balance was so close. Doubling
of the small arable area, for instance, would have changed the picture
greatly, with cattle and their produce leading the way. If Archibald
Stewart had thought of analysing his Cash Book, he might have given
cattle a more prominent place in the work of his year.

The outlay on fodder came to 3.02%. This covered the purchase
and carriage of oilcake, linseed meal, treacle, bran, bruised corn, hay,
straw and some turnips. There was also 'swines meat', for the one or
sometimes two young pigs that were bought each year at an average
cost of 16/3 for fattening and slaughter for domestic consumption,
apart from the one sale in December 1894 of 12 stones of pork at 5/- a
stone. The pig's 'meat' was not cheap, totalling £7.19/- for four years
only of the period covered (1888–91). The total cost of young pigs
bought over thirteen years amounted to only a little over twice that
amount, £18.13.6.

The costs of grazing, i.e. the summering and wintering of sheep
and the summering of stirks outside the farm, have been counted in
with the calculations for sheep and cattle, since they were clearly
identifiable. Ewes and hogs (yearlings) were regularly wintered in
separate flocks, as were small numbers of tups or rams (usually two)
in some years. The ewes and hogs were looked after by R Menzies in
1888, the hogs by D Stewart in 1894 and 1896, the ewes by D
Caithness in 1896, by Robert Campbell in 1897, and Robert Duncan
in 1898. Those who looked after the tups were George Duncan in
1889, Edward Duke in 1893 and Alex Stewart in 1894, 1896 and 1897.
There was no single place to which Archibald Stewart regularly sent
his sheep.

Summering involved lambs only. G Michie looked after them in

1892 and 1893; in 1894 and 1895, John Innes summered them and organised their dipping. Some old sheep were also summered by G Michie in 1893.

Numbers of sheep are given occasionally:

1888: 81 ewes were fed on D. Caithness's turnips for a week
1889: 27 hogs wintered
1890: 80 ewes and 45 hogs wintered
1891: 26 hogs wintered
1892: 41 hogs and 82 ewes wintered
      28 lambs summered
1893: 32 hogs wintered, also 24 other hogs, 87 old sheep, and
      2 rams
1894: 76 lambs summered
1896: 40 hogs wintered
1897: 2 rams wintered.

On this basis, the greatest number mentioned was 145 sheep in 1893. To judge by the sales of lambs, cast ewes, ewe hogs, rams or wethers, and gimmers (yearling ewe hogs after their first shearing), the produce of the flock was just over sixty per annum.

The present-day stock of breeding ewes carried on Buskhead is 120. In 1970, when the Stewarts were still in the place, the flock consisted of Blackfaces: thirty ewes, four rams and forty hogs.

Both food for the house, and fodder for the stock, were got from the millers, David and James Caithness, at the Mill of Aucheen. Purchases included sacks of oatmeal and flour, barley which would have gone into the making of broth, and *dust*, i.e. particles of meal and husk from the grinding that could be used in making *sowens* or flummery.

Besides cereal products, there were also root crops. Half a ton of seed potatoes was paid for in May 1886, but since there is only one other reference to purchase in 1893, it is likely that Buskhead used its own seed for much of the time. Turnip seed was got from John Milne & Sons; yellow turnip seed cost 8d, and Swedish turnip seed, 10d.

Buskhead's own oats were probably being ground in the mill. Seed corn is regularly mentioned at £1 to £1.8/- a quarter, but there is no indication of barley being grown, even though some was supplied by the Caithnesses at times. The fields that had grown oats were laid down to grass in due rotational sequence. For this, grass and clover seed was needed, some of it bought from James Innes at £1.10/- a quarter. Tares are mentioned once only. The outlay of 9d would not have gone very far, so it looks as if a small end-ridge in a field was being planted.

Garden plants and seeds were got from seedsmen like Henderson & Sons, Brechin, and J Lowdon. Peas, parsley, cabbage and 'plants', unspecified, were the main purchases. Three hundred cabbage plants cost half-a-crown in 1891.

Seed potatoes were bought only in 1885 (when half-a-ton cost £1) and in 1893. The Stewarts no doubt used their own seed at other times. Turnip seed, both yellow and Swedish, was bought more regularly, however. Potato and turnip-lifting were jobs that some-times needed extra help, both male and female, for which wages had to be paid, as noted earlier. A turnip-cutter bought for £1 in November 1896 gives a glimpse of the feeding of stock, housed for the winter, on sliced roots. One or (later on) two young pigs were bought each year for fattening. Only towards the end of the period, when the lady of the house was ill, was more use made of shops. In 1898, 11 lbs of tea were bought from Hugh Campbell & Son, at £1.2/-; groceries amounted to £3.7.4; and the flesher's bill, presumably for beef, came to 11/4. The purchase of 20½ lbs dripping in this year, when two young pigs were also bought, is hard to explain.

Meal was no doubt kept in a girnel. On 18 March 1891, Stewart Porter got £3.13.9 for 'Building arc'. This was very likely the meal-ark common throughout Scotland as a food-storage chest.

The food base, therefore, was oatmeal, garden and field vegetables, roots and pork. The half-gallon of whisky bought on 1 December 1886 (8/3) and a further gallon on 31 May 1887 (16/6) would have brightened up special occasions but would hardly have been for daily use.

Draught power was provided by horses, but little information about them can be gleaned. There were regular purchases of nitre as a curative for the horses. One horse was bought in 1897 at a cost of £11.2.3. Amongst the minor purchases were items such as blackening for harness, rope to make plough lines, stable brooms, a curry comb and glasses for the stable lamp. Once, a cart saddle had to be repaired. The outlay on horses was in general quite low.

There are no direct references to farm carts or gigs, though there are a number of entries like 'Expenses of a Man & Horse at Brechin 6/-' (14 March 1888), or 'Expenses of Man & Horse for Load to E. Manse 3/-' (27 May 1889), showing that there could be movement from time to time over quite long distances.

The bills were paid to blacksmiths, as to the joiner, in May and November each year, no doubt for horse-shoeing, the laying of plough socks and coulters, and machinery repairs, though such jobs are not specified. The smith got an average of about £3.7/- a year, and over the period his bill – at 2.69% of total farm expenditure – was over twice as high as for any other craftsman.

More money, in fact, was spent on farm equipment. What is recorded in the Cash Book is fairly miscellaneous in character. The main items were a 'dreel (drill) harrow', 9/6; a fan (winnowing machine) for cleaning threshed grain, supplied by C Wilson & Co, £5; 3 flakes for sheep pens and a harrow from W Black & Son, 15/-; a turnip cutter, £1; a pig trough, 8/6; and a corn bruiser from William Napier, £2.10/-. On 23 April 1891 there is an entry, 'P. Brown for thrashing Mill £13.0.10', and on 12 August 1895, 'P. Brown repairs to mill £2.14.9'. This was Peter Brown, based on the Bucket Mill at Finzean, near Banchory, who built several threshing mills for farmers in the Glen. Since the farm had a good deal of machinery, some of which is indicated above, it is not surprising that bottles of machine oil were often bought.

Looking after the buildings, renewing items of furniture, and keeping the fires going were matters that cost a certain amount from time to time. The roofs were thatched and good-quality wheat straw was sometimes bought for roofing. Six bales were got for £2.2.10 in October 1887, and two more for 18/- in March 1889. Robert Middleton was paid 5/- for two days' work thatching the house in October 1889, a figure well below the cost of the thatch itself. He got 6/- for three more days' thatching in November 1896. Quantities of coir rope were bought from time to time for roping down the thatch.

Work was done on the house, byres and stable by Stewart Porter and William Davidson. This came to £11.14.2 in October 1886. On 5 October 1887, the Cash Book records: 'Paid Stewart Porter for Building Stabel & vent £7.9/-'. It looks as if the work, probably on grounds of cost, was being spread over more than one year.

A big, internal job must have been done in the house in 1887, when George Farquharson, plasterer was paid £6.15.8. In 1890, William Shand repaired the doorstep for 2/-, and in December 1894, Stewart Porter provided a new can to the kitchen 'lum', for 5/-. Bags of cement were got on occasion, and once, in June 1892, there is an entry for two dozen bricks at 3/6. The purpose is not stated, but no doubt some work was being done to a fireplace.

Indirect glimpses of the interior of the house can also be got from the brief entries. Walls were plastered, as noted above, and some were papered, for William Bruce, paperhanger, was paid 11/8 in November 1889. H. Black & Son provided 'Floorcloth' at £3.8/- in 1888, for new floor covering. Six chairs were bought in 1887 for £2.8/-, a mahogany cabinet from H. Black & Son in 1889 for £6, an iron bed with mattresses in the same year for £2.7.6, and a washing tub in 1895 for 6/3.

Heating and cooking were done with open fires. By this date, coal was being bought regularly. A ton cost £1.6.10 in October 1895, and a bag was 2/3 in July 1898. It was ordered and delivered and involved no work. Peat, on the other hand was labour intensive, often requiring extra help that had to be paid, as the wages bills show. It is little wonder, therefore, that coal had come to be the preferred fuel, with peat being used as a supplement, as indicated by proportional costs for the period covered: coal £11.1.7, peat £1.15.0.

In the yearly round of work, cattle had their own rhythms. Two and sometimes three cows were served regularly at prices ranging from 4/- to 5/-, and produced calves in due course. Stirks were sold in Brechin when they were ready. Just as sheep were often wintered off Buskhead, so were stirks summered at other places, for example three with J Pirie at 1/6 per head per week in 1893 and again in 1894. In 1896 and 1898, three stirks were summered with Archibald Stewart, which presumably means they were kept on Buskhead.

Stock was bought now and again. A 'quay' (heifer) cost £10 in 1886, and another £8.10/- in 1888. A cow was £13 in 1888, and one in calf was £14 in 1898. Alongside the annual net contribution of 28.7% to the farm income, the milk, butter and cheese would have been a not insubstantial addition.

Cattle in the byres, horses in the stable, the pigs in their sties and the poultry in their houses produced muck which in due course was spread on the fields and in the turnip and potato drills. All the same, lime and artificial manure were also bought in fair quantities that amounted to 3.35% of the expenditure over the period. The lime supplier was John Hood (who also supplied coal); manure came from Charles Tennat & Co, probably in Edzell, John Milne & Co, and the Brechin Agricultural Company. Lime was bought at around 5/9 a boll, and manure at 6/- to 7/- a hundredweight. It seems that by this date the circular stone lime kilns to be found at various places in the Glen – e.g. at Ardoch, Tirlybirly, Whigginton and Dykeneuk – had gone out of use.

The fields were kept free of moles by William Glass, whose attentions cost 1/- in 1889, and rose to 3/- in 1898. For the most part the fields were surrounded by post and wire fences. Wire was bought from time to time in hundredweight hanks, and larch posts were bought on a number of occasions from William Black & Son, some of them for a march fence (14 Dec. 1887). Eight straining posts were got for 4/- in May 1888, for taking the pull of the wire at field corners and gate openings. Staples were used to fix the wire to the posts. Dykes are mentioned only once, in October 1887, when R Middleton received £1 for building one. This may have been the garden dyke.

Though the entries are terse, analysis can nevertheless produce a great deal of information on the economics and the life and work of Buskhead towards the end of the nineteenth century. The pattern would not have been much different on the other farms in the Glen.

W A Stewart, born at Buskhead in 1902, has provided an outline of the daily outline around the period of the First World War. Rising time was 5–6 a.m., when the men fed the horses, and the women went to the byre. Breakfast was before 7 a.m., then the men went out to work, until 11 a.m. Dinner was at 11.15 a.m., and work started again at 1 p.m., lasting until 5.30 to 6 p.m., with mid-yokin' at 3 p.m. during harvest, etc. The horses were fed before the men, fed again between 8 and 9 p.m., and bedded down for the night.

The seasonal routine was as follows:

January-February	Ploughing. Got the dung out.
March	Ploughing. Started to sow.
April	Finishing sowing. Lambing.
May	Got in potatoes and turnips. Lambing.
June	Odd jobs. Cutting peats. Cutting hay late in the month.
July	Getting in the hay. Thinning the turnips.
August	Harvest. Men who could be spared might go to the hill to the shooting. Sheep sales.
September	Sales of stirks.
October	Potatoes towards the middle of the month. Grubbing stubble. Sometimes the harvest continued into November or later.
November	Much the same, and got in turnips.
December	Stubble ploughing.

These details can be used to flesh out the bones of the Cash Book, since there were few fundamental changes in the work as long as horses provided the main source of draught.

Note

Egg-production and income from eggs are recorded in later pen-
cilled entries in the Cash-Book. Since eggs could play a role in the
domestic side of the economy, providing spending money for the
women, the figures are worth noting. The year 1937 gives the most
complete detail:

Year	Month	No. of Eggs Sold (Dozens)	Price
			£ s. d.
1937	Jan.	22	1. 9. 8
	Feb.	28½	1.17. 3
	Mar.	62½	2.17. 3
	Apr.	43	1.16. 2
	May	43½	1.16. 3
	June	53½	2.12. 4½
	July	54½	3. 5. 9
	Aug.	46	3. 7. 5
	Sept.	34½	2.12. 6
	Oct.	19½	1.13. 7
	Nov.	17½	1.12. 4
	Dec.	18	1.18. 2

This gives a total of 443 dozen, bringing in a sum of £26.18/8 at an
average price of ½d a dozen.

TEXT OF
ARCHIBALD STEWART'S
BUSKHEAD CASH BOOK

The following pages give an accurate transcription of Archibald Stewart's Cash Book, printed with the Debit side on the left-hand page and the Credit side on the right.

1885		Stock Dr to Cash	£	S	D
Oct	10	Money on hand	35	"	"
"	"	In Bank	11	"	"
"	16	B Gibb for Lamb	8	8	"
1886					
Jan	13	Sold 1 Cow	7	15	6
May	19	Sold 3 Stirks	23	3	9
			85	7	3
			85	7	3

1885		Contra Cr	£	S	D
Oct	10	To Cash Book	"	2	6
"	16	Paid Croll & Will	4	"	"
"	"	Mitchell & Son Brechin	1	11	6
"	"	John Hood "	7	9	1
"	"	Black & Johnston	"	8	7
"	"	Sheep dip	"	8	"
"	"	Paid W Davidson	4	4	"
"	"	Funeral expenses	2	14	5
"	"	D & I Caithness Mill Aucheen	"	19	"
"	19	D Moir Blacksmith	"	14	
"	26	1 Bottle of Machine oil	"	1	"
"	"	2 Cows Served Mr Michie	"	10	"
"	"	1 " " Mr Campbell	"	4	"
"	"	Stewart Porter Making Grave	"	6	"
"	"	D Caithness 2 Rams at 30/- each	3	"	"
Novr	24	Rent of Buskhead for Six Months	7	10	–
"	"	C Crow wages	7	–	–
"	"	I Gold "	2	–	–
"	"	I Young Seedsman	–	2	9
"	"	Poor Rates	–	7	3
"	"	engaging Servant	–	1	"
			43	13	1

1886	Stock Dr to Cash	£	S	D
	Brought Forward	85	7	3

1885		Contra Cr	£	S	D
		Brought Forward	43	13	1
Dec	1	Carriage of oil Cake & Bran	"	3	"
"	8	Paid 1 St Barley	"	2	"
"	"	Sheep dip	"	10	8
"	"	Road Money	"	4	8½
"	31	Oil Cake & Bran to I Young	1	6	9
1886					
Jan	13	Expenses of Cow to Sail	"	7	"
Feby	2	Fire Insurance	"	15	"
"	18	Niter for Horses	"	2	"
"	"	1 Bottle of Machien oil	"	1	"
March	25	Carriage of oil Cake	"	1	"
April	3	Paid 1 Cwt Lintseed Cake	"	10	"
"	"	expenses at Brechin with Horse & Cart	"	3	"
April	10	Paid carriage of 1 load of Straw to Edzell	"	4	"
"	22	expense going for Hay	"	2	"
May	6	to ½ ton Seed Potatoes	1	"	"
"	11	1 boll meal & 1 Sack dust	"	19	"
"	19	expenses at Brechin with Stirks	"	7	"
"	20	Paid wintering Ewes	11	"	11
"	"	Carriage of Bran	"	1	"
"	22	44 Stones Hay at 10d per St	1	16	8
			63	9	9½

1886		Stock Dr to Cash	£	S	D
		Brought Forward	85	7	3
May	26	Reduction rent	1	10	"
"	"	Sold 2½ quarters Grass Seed	2	7	6
Augt	16	Price of Wool	10	6	"
"	"	Sold 12 Ewe Lambs at 15/3 each	9	3	"
"	"	20 weather Lambs at 13/3 each	13	5	"
Oct	12	20 Cast Ewes at 19/ each les 5/	18	15	"
			140	13	9

| | | | £140 | 13 | 9 |

1886		Contra Cr	£	S	D
		Brought Forward	63	9	9½
May	26	I Christison Wages	5	10	"
"	"	D Caithness 114 Stones 12 lbs Hay at 1/			
		per St	5	14	6
"	"	Blacksmiths A/C	1	14	11
"	"	engaging Servants	"	2	"
"	"	I Lowdon for plants for Garden	"	2	"
"	"	Six Months Rent of Buskhead	7	10	"
June	3	1 Cwt oil Cake & Carriage	"	11	"
"	10	10½ Cwts Manure	3	2	"
"	"	10 Cwts 14 lb Straw	1	11	6
"	"	Carriage to Edzell	"	2	6
Aug	23	I Young Seedsman A/C	4	4	7
"	"	Paid Young Pig	"	17	"
Oct	12	D & I Caithness Mill Aucheen	3	16	"
"	"	B M Bisset Brechin	"	2	"
"	"	H Peter 5 Bales Straw at 6/9 each	1	13	9
"	"	Stewart Porter a/c for House & Bires	5	2	4
"	"	Wm Davidsons for House & Bires	6	11	10
"	"	two Checques Cashed	"	1	"
"	"	Bought 1 quay	10	"	"
"	"	W Glass for the Harvest	"	12	"
			122	10	8½

1886	Stock Dr to Cash	£	S	D
	Brought Forward	140	13	9
		£140	13	9

1886		Contra Cr	£	S	D
		Brought Forward	122	10	8½
Nov	22	1 Can of Sheep dip	"	2	6
"	"	Six Months rent of Buskhead	7	10	"
"	"	Jane Cristison wages	6	"	"
"	"	Alex Middleton do	1	13	"
"	"	to engaging Servant	"	2	"
"	"	Blacksmith A/C	1	16	4
"	"	G Michie for 2 Cows served	"	10	"
"	"	Mr Campbell 1 do do	"	5	"
Dec	1	Poor & School rates	"	7	3
"	"	½ Gallon Whiskey to C Mitchell	"	8	3
"	"	Niter for Horses	"	2	"
"	"	expenses of Rams coming Home	"	2	7
"	"	Road money	"	3	9
"	28	expenses of 2 Rams down to Arbroath	"	1	"
1887					
Jan	31	Carriage of oil Cake & bran	"	5	"
Feby	3	Paid George Farquharson Plasterer	6	15	8
"	"	Gave Mother to pay her A/C	5	3	"
April	4	Paid Carriage of seed Potatoes	"	2	"
May	18	W Glass for Moles	"	4	"
"	"	J Sutherland Pasture	1	"	"
			155	4	"½

1887		Stock Dr to Cash	£	S	D
		Brought forward	140	13	9
May	3	Sold two Heifers	22	6	6
"	18	Sold two one year old Stots	14	"	"
"	"	to Keep for Horse for winter of			
		1885 & 1886	5	"	"
"	"	Sold one Ram	1	5	"
Augt	5	To price of Wool & 2 Skins	12	9	"
"	13	Sold 20 weather Lambs at 10/3 each	10	5	"
"	"	" 20 Ewe Lambs at 12/3 "	12	5	"
Oct	5	Sold 23 Cast Ewes at 18/3 "	20	19	9
"	"	to Keep for Horse for winter of			
		1886 & 1889	5	"	"
			244	4	"

1887		Contra Cr	£	S	D
		Brought Forward	155	4	"½
May	18	Paid Ewes Wintering	9	12	9
"	"	expenses of Hogs from Brechin	"	1	6
May	26	Blacksmiths a/c	3	"	2
"	"	Wages to R Burness	5	"	"
"	"	engaging Servants	"	2	"
"	"	Six Months rent of Buskhead	7	10	"
"	31	1 Gallon Whiskey to C Mitchell	"	16	6
"	"	John Hood for Lime	4	18	3
"	"	expenses at Brechin at Sail	"	7	"
"	"	Paid young pig	"	16	6
"	"	Paid J Perie for 40 Stones Hay and 2 lodes Turnips	2	6	8
"	"	R Duncan 1 Hank Wire	"	9	6
Augt	11	Paid J Corral Carriage wool Bags, rope & dip	"	2	"
Sept	20	Paid A Bowman for Harvest work	"	8	"
Oct	25	6 Bales wheat Straw to H Peter	2	2	10
"	"	Paid W Black & Son a/c	5	3	4
"	"	Brechin Agricultural & Trading Co	2	13	"
"	"	Paid James Young Seedsman	3	1	7
"	"	C Mitchell & Son Brechin	"	16	6
"	"	B M Bisset VS	"	2	6
		Carrie. Forward	£204	14	7½

1887	Stock Dr to Cash	£	S	D
	Brought forward	244	4	"
		£244	4	"

1887		Contra Cr	£	S	D
		Brought Forward	204	14	7½
Oct	5	Paid Stewart Porter for Building Stabel			
		& vents	7	9	"
"	"	Paid Commison for Ewes at Brechin Sail	"	7	"
"	28	R Middleton for Building dyke	1	"	"
Nov	22	R Burness wages	5	10	"
"	"	A Middleton wages	1	15	"
"	"	Rent of Buskhead Half year	7	10	"
"	"	Wm Davidson Wright a/c	4	"	8
"	"	D Moir Blacksmith a/c	"	17	11
"	"	Ritchie Sheep Dip	"	10	6
"	"	Poor & School rates	"	7	3
"	"	Repairs to Halters	"	1	"
"	"	Engaging Servants	"	2	"
"	30	D & I Caithness Millers	1	6	"
Dec	14	Road Money	"	3	9
"	"	Niter for Horses	"	2	"
"	"	Paid Half of a 100 posts for march Fence	"	8	4
"	"	Paid H Johnston 3/ for carting Thatch to			
		Edzel	"	3	"
1888					
Jany	14	Expense of 1 Barrel of Tracle to Edzell	"	2	10
"	"	Bought 6 Chairs at 8/ each	2	8	"
"	"	Dalhousie Mamorial	1	"	"
			239	18	10½

1888		Stock Dr to Cash	£	S	D
April		Brought Forward	244	4	"
"	3	Sold 1 Cow at Brechin Auction Mert	13	17	6
May	1	Sold 3 Stirks at Brechin Auction Mert			
		at £7 17 6 each	23	12	6
June	3	Sold 3 qtr Grass Seed	3	"	"
"	"	Wintering of Horse	5	"	"
Augt	4	Sold 432 lbs Wool at 5½d per lb	9	18	"
"	11	Sold 40 Weather Lambs at 11/9 each	23	10	"
"	"	10 Ewe Lambs at 12/6 each	6	5	"
Oct	11	sold 22 Ewes at 20/3 each	22	5	6
"	"	1 Weather	1	6	"
"	"	Sold 2 Rams at 22/6 each	2	5	"
			£355	3	6

1888		Contra Cr	£	S	D
		Brought forward	239	18	10½
Jany	14	Carriage of Chairs	"	2	1
Feby	2	Fire Inshurance	"	15	"
March	14	1 Sack of Bran	"	6	"
"	"	2 Cwts of Sement	"	6	"
"	"	1 Bottle of Machien Oil	"	1	"
"	"	Expenses of a Man & Horse at Brechin	"	6	"
April	3	Keep expenses & commison of Cow at Sail	"	12	6
"	7	To D & I Caithness ½ boll of flour 8/- brusing 1 qu Corn 4/-	"	12	"
May	1	Keep expenses & Commison of Stirks at Sail	"	16	9
"	"	Bought 1 young pig	"	16	"
"	"	Carriage of pig & 1 Cwt of Oilcake	"	1	6
"	"	Paid Agricultural Co	3	7	"
"	"	Paid H Peter 2 Bales Straw	"	15	6
"	"	8 Straning Posts	"	4	"
"	"	J Peerie 51 Stones Hay at 1/- per St	2	11	"
"	23	1 lb yellow Turnip seed	"	"	8
"	24	Blacksmiths a/c	1	6	9
"	"	R Menzies for Wintering Hogs & Ewes	19	1	4
"	"	D Caithness 1 week Turnips to ewes 2½ per Head		16	10½
"	26	R Burness wages	5	10	"
			£278	6	10

1888	Stock Dr to Cash	£	S	D
	Brought Forward	355	3	6
		£355	3	6

1888		Contra Cr	£	S	D
		Brought forward	278	6	10
May	26	Six Months Rent of Buskhead	7	10	"
"	"	Paid Share of Shepherd's Wages herding			
		the Ewes		18	6
June	14	To Black & Son for fence in front of			
		House	1	6	2
"	"	Henderson & Son Seedsmen	1	2	2
"	"	C Mitchell & Sons	"	8	3
"	"	Brechin Manure Co	"	14	6
"	"	B M Bisset	"	1	6
"	16	Bought a dreel Harrow	"	9	6
Augt	14	D & J Caithness for Meal & Swines meat	2	7	6
Sept	22	Bought one Cow	13	"	"
Oct	3	D & J Caithness for Meal & Swines meat	2	2	"
"	15	Bought one Quay	8	10	"
"	16	Wm Napier 30 lbs Coir Rope & Carriage		6	6
Nov	3	C Mitchell & Son		16	6
Oct	11	Commission & expenses of Ewes at sail		8	6
Novr	"	Black & Son Floorcloth	3	8	"
"	"	Sheep Dip	"	9	"
"	"	3 days of a man at the harvest at 2/6 per			
		day	"	7	6
"	"	Nails	"	2	"
"	"	Niter for Horses		2	6
			322	17	5

1888		Stock Dr to Cash	£	S	D
		Brought Forward	355	3	6

1889					
Feby	2	Income up to this date	£355	3	6
		Expenditure up to this date Deducted	343	2 11½	
		Fuinds on hand at this date	£12	0	6½

Archd Stewart

1888		Contra Cr	£	S	D
		Brought Forward	322	17	5
Novr	21	Paid Blacksmiths a/c	1	7	3
"	22	R Burness wages	6	"	"
"	"	A Middleton	2	"	"
"	"	Six Months rent of Buskhead	7	10	"
"	"	Engaging Servants	"	2	"
"	24	Bought one Ram	1	10	"
Dec	11	Road Money	"	3	9
"	"	½ Cwt Linseed Cake	"	4	6
"	20	School & Poor Rates	"	7	3
"	"	Expense of one Ram from Perth	"	1	3½
1889					
Jany	2	Expenses of two Rams to George Duncan	"	1	"
"	14	W Glass for Moles	"	1	"
"	18	Shaft to Axe	"	"	6
"	22	W Shand repairs to House	"	2	"
Feby	2	Fire Insurance		15	"
			343	2	11½

1889		Stock Dr to Cash	£	S	D
		Ballance brought forward	12	0	6½
March	19	Sold 1 Cow Brechin Sail	12	10	"
May	1	Sold 1 Heifer £11–10 2 Stirks a £9 each	29	10	"
"	29	Sold 6 Quarters Corn at 18/ per Quarter	5	8	"
July	16	Sold 3½ Quarters Grasseed at 15/ per			
		Quarter	2	12	6
Aug	2	Sold 547 lbs wool	12	14	6
"	12	Sold 32 Lambs at 13/6 each	21	12	"
Oct	9	Sold 27 Cast Ewes & 2 Weathers at 24/ each			
		les 10/ back	34	6	"
"	14	Sold 1 Ferow Cow	9	15	"
"	"	Wintering of 1 Horse to Lord Hindlip	5	"	"
1890			£145	8	6½
Jany	22	Sold 3½ Quarters Corn at 17/6 per			
		Quarter	3	1	3
			£148	9	9½

1889		Contra Cr	£	S	D
March	19	2 Bales Straw for Thatch H Peter	"	18	"
"	22	W Black & Son Mahogany Cabnet	6	"	"
"	19	Expenses & Commission of Cow to Brechin	"	10	"
"	22	to Expenses of man & horse with Grass seed	"	7	"
May	1	Expenses of Cattle of Sail & Commission	"	16	"
"	6	Wintering of 27 Hogs at 9/7 each	12	18	9
"	27	Bought one young Pig	"	13	6
"	"	Rachel Burness Wages	6	"	"
"	"	Six Months rent of Buskhead	7	10	"
"	"	Blacksmiths A/C	1	13	11
"	"	Wm Davidson Wright	"	5	8
"	"	3 quarters seed Corn at £1 per quarter	3	"	"
"	"	C Mitchell & Son	"	16	6
"	"	2 St Linseed Meal	"	4	8
"	"	Carriage of 1½ Cwts Manure	"	1	1
"	"	Expenses of Man & Horse for Load to E Manse	"	3	"
"	"	1 Sack Flour 3 qu Corn Cut & 2 bushels Barley	1	7	"
"	"	B M Bisset VS	"	1	"
"	"	Wintering of Ewes	10	"	"
June	10	The Brechin Agricultural Co	4	7	6
July	16	Henderson & Sons Seedsmen Brechin	1	8	3
			£59	1	10

Stock Dr to Cash	£	S	D
Brought Forward	148	9	9½

1889		Contra Cr	£	S	D
		Brought Forward	59	1	10
July	16	Bought 1 young Pig		15	6
"	"	Tar	"	2	"
"	"	Steples for fence	"	2	"
"	"	1 Sack of meal & pigs meat	1	"	"
"	30	Bought 1 Iron Bed with matresses	2	7	6
"	"	Nails	"	1	"
"	"	Expenses to Brechin with Wool	"	3	6
Augt	17	1 Bag of Pigs meat	"	8	3
"	20	Wash for Lambs	"	2	8
"	28	1 Bag of Pigs meat	"	8	"
Sept	18	Do Do	"	8	"
Oct	14	Do Do	"	8	"
"	16	Bought 1 Ram	1	10	"
"	"	Nails	"	"	6
"	18	W Black & Son Brechin	2	18	1½
"	"	C Mitchell & Son	"	8	3
"	25	R Middleton 2 days thatching House	"	5	"
Nov	1	Wm Bruce paperhanger	"	11	8
"	"	Cheque Book	"	1	"
"	9	Blacksmiths a/c	3	12	9
			£74	15	6½

1890		Stock Dr to Cash	£	S	D
Feby	17	Brought Forward	148	9	9½

	£	S	D
Feby 17 Income up to this date	148	9	9½
Expenditure up to date deducted	102	3	4½
Fuinds on hand at this date	£46	6	5

Archd Stewart

1989		Contra Cr	£	S	D
		Brought forward	74	15	6½
Nov	19	Paid George Michie 2 Cows serving at			
		4/ each	"	8	"
"	21	Wm Davidson Wright	"	14	7
"	22	Ratchel Burness wages	6	"	"
"	"	Alex Middleton Do	2	10	"
"	"	Earl Dalhousie Six months rent	7	10	"
"	"	James Caithness Mill of Aucheen 1 cwts			
		Barley	"	14	"
"	"	Alex Jack Sheep Dip	"	7	6
"	"	Expenses to Edzell at this Term	"	2	6
"	25	John Hood Straw for thatch & Coal	3	18	10
"	"	Charles Tennat & Co Manure	"	11	3
"	29	to Glasses for Stable Lamp	"	2	"
"	"	Bought one Ram	1	10	"
Dec	3	Wm Napier 18 lbs Coir Rope	"	4	6
"	9	School & Poor Rates	"	7	7
"	"	Road Rates	"	4	1
"	17	Blackning for Harness	"	1	6
1890					
Janr	22	Wm Shand reparing door step	"	2	"
"	"	Bought 100 posts for March Fenses	"	18	9
Feby	8	1 Bottle Oil	"	1	"
"	11	Carriage of 2 Cwts oil Cake		1	6
"	"	Fire Insurance		18	3
			£102	3	4½

1890			£	S	D
		Stock Dr to Cash			
		Brought Forward	46	6	5
April	9	Sold 10 bushels Grass seed at 2/6 per bu	1	5	"
"	11	Sold 16 Do. D. at 2/6 per bu	2	"	"
May	29	Sold 2 one year old Bullocks at			
		£10/10 each	21	"	"
June	12	Wintering of Horse	5	"	"
July	28	Stewart Bro for wool £15 9/ Less 6d for			
		Cheque	15	8	6
Augt	13	Sold 27 Weather Lambs at 14/6 each	19	11	6
Oct	9	Sold 27 Ewes at 24/6 each Less 1/6	33	"	"
"	"	10 Gimmers at 32/3 " Less			
		Commission 4/	15	18	6
Nov	22	Sold 2 Rams at 22/ each	2	4	"
"	"	Sold 3 quarters Corn 17/6 per qu	2	12	6
1891					
Feby	15	Income up to date	£164	6	5
		Expenditure deducted	119	16	10½
		Funds on hand at date	£44	9	6½
		Feby 15th 1891			
		Archd Stewart			

38

1890		Contra Cr	£	S	D
Feby	18	Comb for Horses 6d Nails 6d	"	1	"
"	25	Cwt Bran	"	6	"
March	10	Carriage of Cake	"	1	6
"	18	A. Birse 1 day's work	"	2	
"					
"	23	Meadicine for Cow	"	1	"
"	24	Bought 2 Quarters Corn for seed at 18/6			
		per Quarter	1	17	"
April	2	2 young pigs at 16/6 each	1	13	"
"	"	to 3 days work at Grass seed	"	4	6
"	22	peas & persly &c for Garden	"	"	7
"	"	1 Bottle of Machien Oil	"	1	"
"	"	Edzell Railway	1	"	"
"	25	George Duncan 1 day	"	1	"
"	"	Bran 6/6 Cabbage 6d	"	7	"
May	6	Carriage of Cake	"	"	10
"	15	Expenses at Edzell for Manure	"	2	6
"	"	repairs to Cart Saddle	"	1	"
"	26	A Middleton wages	"	10	"
"	"	A Low "	"	5	10
"	26	Engaging Servants	"	3	"
"	"	Wintering of 80 Ewes	8	13	4
"	"	Do of 45 Hogs	22	10	"
			£43	6	9

1890		Contra Cr	£	S	D
		Brought forward	43	6	9
May	26	Earl Dalhousie 6 Months rent	7	10	"
"	"	Expenses at Term Market	"	2	"
"	"	Niter for Horses	"	2	10
"	"	Wm Glass Moles	"	2	"
"	"	D Moir Blacksmith	"	17	6
June	7	Alex Stewart	28	18	10
"	12	Brechin Agricultural Co	4	"	"
"	"	Wm Napier Ironmonger	"	9	11
"	"	C Mitchell & Son	"	16	6
"	"	1 Fan from C Wilson & Co	5	"	"
July	15	Expenses at wool sail	"	5	"
"	26	Carriage of Fan 1/4 Expense of wool of Brechin 3/	"	4	4
"	"	Bought 1 Shovel	"	2	6
Augt	3	to 3 days work at 2/ per day	"	6	"
"	18	3 Gallons of pitch Oil	"	2	"
"	"	1 day of a man	"	1	6
Sept	15	to 1 Sack of pigs meat	"	8	6
Oct	9	Bought 2 Rams at 21/6 each	2	3	"
"	"	Expenses at sail	"	6	"
July	15	Henderson & Sons Seedsmen	1	2	6
			£96	7	8

1890		Contra Cr	£	S	D
		Brought forward	96	7	8
Oct	18	C Mitchell & Son	"	16	6
"	"	J Hood 5 Bolls Lime with Cartage	1	9	"
Nov	18	1 Sack of Bran with Carriage	"	6	9
"	"	Wm Napier 3 Balls Coir Rope	"	5	3½
"	21	David Moir Blacksmith	1	2	9
"	22	Tutors of the Earl Dalhousie	7	10	"
"	"	Ann Low wages & engaging for Six months	6	2	"
"	"	John Valentine wages	2	"	"
"	"	Dr Campbell 1 Cow served	"	5	"
"	"	to expenses at Market	"	3	"
Dec	4	Poor & School Rates	"	7	7
"	"	County Rates	"	5	10
"	"	B M Bisset	"	18	"
"	15	R Middleton 8 days work	"	13	"
1891					
Jany	5	Oil Can	"	1	6
"	"	Nails	"	2	6
"	17	Carriage Cake 3 Cwts at 10d each	"	2	6
Feby	2	Fire Insurance	"	18	"
			£119	16	10½

1891		Stock Dr to Cash	£	S	D
March	18	Sold 1 Fat Heifer	20	"	"
		Brought Forward	44	9	6½
May	20	Sold two stirks at £8.10 each	17	"	"
"	"	12 bushels Grass seed at 3/ per bu	1	16	"
June	19	Wintering of Horse	5	"	"
Augt	14	Sold 580 lbs wool at 6d per lb less 5/	14	5	"
"	"	Sold 30 Lambs at 11/ each	16	10	"
Oct	8	Sold 24 Ewes at 17/ each	20	8	"
"	"	" 3 Weathers at £1.5/9 each	3	17	3
"	20	" 11½ Quarters Corn at £1 2/ per Qu	12	13	"
			£155	18	9½

1891		Contra Cr	£	S	D
Feby	21	Bought 1 Horse	29	"	"
"	"	Traveling expenses	"	5	"
March	16	2 qrs Seed Oats at 18/6 per qu	1	17	"
"	18	Stewart Porter Building Arc	3	13	9
April	21	Bought 2 pigs at 14/6 each	1	9	"
"	"	Carriage of Cake 3 Cwts at 9d each	"	2	3
"	23	P Brown for thrashing Mill	13	"	10
May	2	G Duncan 12 days Lambing Ewes	"	12	"
"	3	1 Sack of Bran	"	7	3
"	"	Garden Seeds	"	1	1½
"	"	300 Cabbage plants	"	2	6
"	"	1 Bag of Bran	"	7	6
"	"	Carriage of Cake Bran &c	"	1	4
"	23	1 Bottle Machien Oil	"	"	8
"	"	To Wintering of 26 Hogs at 7/1 each	9	5	9½
"	"	David Moir Blacksmith A/C	2	4	6
"	"	David Mitchell Joiner	1	17	7
"	24	Dr Ironside	2	11	"
"	28	Ann Low wages	6	"	"
"	"	Tutors of the Earl Dalhousie	7	10	"
"	"	Engaging Servants	"	2	"
			£80	11	1

Stock Dr to Cash	£	S	D
Brought Forward	155	18	9½

	£155	18	9½

1891		Contra Cr	£	S	D
		Brought Forward	80	11	1
May	28	John Hood Coal & Cartage	1	2	2
"	"	C Mitchell & Son	1	13	"
"	"	W Black & Son	1	16	8½
June	19	Wintering of Ewes	10	2	6
"	"	Wm Glass Moles	"	2	"
July	1	A Birse 2 days Turnip Howing	"	4	"
"	"	I Caithness 1 Sack of Pigs meat	"	11	4
"	"	Carriage of Manure & Expenses to Edzell	"	4	6
"	"	28 lbs Lintssed Meal	"	7	6
"	"	Sheep Dip	"	3	11
Sept	1	Brechin Agricultural Co for cake	2	11	6
"	"	1 Bag of Cement & Carriage	"	7	6
Augt	11	G Duncan 2 days at Peats	"	3	"
Sept	2	7 days of R Middleton at harvest 2/ per day	"	14	"
"	"	Ropes for Stacks	"	4	"
"	"	Dip for Lambs	"	4	8
Oct	8	1 Bag of Swines Meat	"	9	2
"	"	Expenses at Brechin Sail	"	5	"
"	"	Bought 1 Ram	1	"	"
"	28	G Duncan 2 days at Potato lifting	"	2	"
			£102	19	6½

	£	S	D
Stock Dr to Cash			
Brought Forward	155	18	9½

	£155	18	9½

1891		Contra Cr	£	S	D
		Brought Forward	102	19	6½
Oct	31	John Milne & Co 10½ cwts Manure	3	17	"
Nov	3	John Hood Coal & Lime	3	13	3
"	"	Henderson & Son Seedsmen	1	5	"
"	"	Wm Black & Son 30 Larch posts	"	10	"
"	"	C Mitchell & Son	"	16	6
"	"	Bought Rope for plough lines	"	2	"
"	5	Oil & Nails	"	1	"
"	24	David Moir Blacksmith	1	1	2½
"	"	Poor & School Rates	"	7	7
"	28	Ann Low Wages	6	"	"
"	"	Even McIntosh	2	"	"
"	"	Six Months Rent of Buskhead	7	10	"
"	30	Wm Davidson Wright	"	16	"
Dec	2	Carriage of Cake & Bran	"	1	6
"	21	to one Stable Broom	"	2	"
"	"	Sheep Dip	"	4	6
"	30	Mrs Campbell to cows serving	"	10	"
1892					
Jany	13	to carriage of Cake	"	"	9
"	26	1 Broom & Carriage of Cake	"	3	3
Feby	8	1 Curry comb	"	1	"
			£132	2	1

1892		Stock Dr to Cash	£	S	D
		Brought Forward	155	18	9½
Feby	23	Income up to date	£155	18	9½
		Expenditure deducted	133	2	3
		Funds on hand at date	£22	16	6½
		Feby 23rd 1892			

Archd Stewart

1892		Contra Cr	£	S	D
		Brought Forward	132	2	1
Feby	9	Fire Insurance	"	18	"
"	16	Carriage of Cake	"	"	9
"	23	1 Bottle oil 8d Carriage of Cake 9d	"	1	5
			£133	2	3

1892		Stock Dr to Cash	£	S	D
		Brought forward	22	16	6½
March	23	Sold 1 Fat Heifer	18	"	"
April	15	To 1½ Quarters Grass Seed at 26/ per Qu 1		19	"
June	8	To 7 Quarters Corn at 22/ per Qu	7	14	"
"	"	To wintering of Horse for Lord Hindlip 5		"	"
July	25	To 559 lbs wool at 5d per lb	12	10	"
Augt	1	To £1 1/ of an overcharge returned by			
		Alexr	1	1	"
Sept	13	Sold 5 Ewes at 10/9 less commission 9d	2	13	"
Oct	13	20 Ewes at 12/8 do at 10/6 less commission			
		4/4	16	"	"
			£87	13	6½

1892		Contra Cr	£	S	D
March	8	Sheep Dip	"	4	8
"	"	Carriage of Cake	"	"	9
"	22	do do	"	"	9
"	29	do do	"	"	9
April	5	do Clover seed	"	"	3
"	7	R Middleton 1 day reparing Houses	"	2	"
"	12	Bought 2 pigs at 17/6 each	1	15	"
"	"	plants for Garden &c	"	4	1
"	"	Carriage of Cake	"	"	9
"	19	do do	"	"	9
"	26	do Bran	"	"	9
"	30	George Duncan 1 day at Garden	"	2	"
May	14	1 lbs Sweedish Turnip seed	"	"	10
"	17	Expenses at Brechin for Manure	"	4	6
"	21	R Middleton 4½ days at Turnips	"	10	6
"	28	Tares 9d 1 Bottle Oil 1/	"	1	9
"	"	Even Tosh for Lambing Ewes & work	1	10	"
"	"	Ann Low wages	6	10	"
"	"	Engaging Servants	"	2	"
"	"	to wintering of 41 Hogs at 7/7 each	15	10	11
"	"	to wintering of 82 Ewes	10	15	3
			£37	8	3

1892		Stock Dr to Cash	£	S	D
Dec	17	Brought forward	87	13	6½

£87   13 6½

1892		Contra Cr	£	S	D
		Brought Forward	37	8	3
May	28	6 Months rent of Buskhead	7	10	"
"	"	Thomson Saddler	"	1	9
"	30	D Moir Blacksmith	1	10	8
June	2	Archd Stewart wages from 15 Oct to			
		2 June 33 week a 10/ per week	16	10	"
"	7	Agricultural Co Manure & Cake	7	7	"
"	"	To 1 Cwt of wire & Stiples	"	11	9
"	8	W Glass Moles	"	2	6
"	"	Nails	"	"	6
"	"	To 3 days of D Ross at peats at 2/6 per			
		day	"	7	6
"	"	" 2 doz Bricks	"	3	6
"	"	" ½ Cwt Linseed meal	"	5	6
"	16	Tar for Sheep	"	3	6
July	19	C Mitchell & Son	1	13	"
"	"	Alexr Bairner 1 Gallon Whiskey	"	17	"
"	"	Expenses at Wool Sail	"	3	"
"	23	George Duncan 2 days building peats	"	3	"
"	"	Carriage of wool Bags & Skins	"	"	7
"	25	Expenses of wool to Brechin Station	"	2	4
June	10	David Mitchell Joiner	"	11	8
Sept	12	Henderson & Sons Seedsmen	1	10	"
			£77	3	"

1892		Stock Dr to Cash	£	S	D
Dec	17	Brought forward	87	13	6½
"	"	Skins		6	

| | | | £87 | 19 | 6½ |

1892		Contra Cr	£	S	D
		Brought Forward	77	3	"
Sept	13	Expenses at Sail	"	4	"
Oct	8	Bought 1 Ram	1	"	"
"	13	Expenses at Sail	"	4	"
"	15	Nails & Oil	"	2	"
"	24	Summering of 28 lambs to G Michie	2	15	2
Nov	11	2 Stirk Summerd to J Peirie	3	16	8
"	28	Annie Low wages	6	10	"
"	"	Even McIntosh do	3	10	"
"	"	6 Months Rent of Buskhead	7	10	"
"	"	D Moir Blacksmith	"	18	10½
"	29	C Mitchell & Son	"	8	3
"	"	Engaging Servants	"	2	"
"	30	D & I Caithness Millers	3	14	8
Dec	3	Poor & School Rates	"	7	7
"	"	3 Cows Serving at 4/ each to Dr Campbell	"	12	"
"	9	Dr Ironside	1	18	"
"	"	Niter for Horses	"	2	6
"	"	Coir Rope for Houses	"	5	"
"	17	County Rates	"	4	11½
"	19	Sheep Dip	"	15	"
			£112	3	8

Stock Dr to Cash	£	S	D
Brought forward	87	19	6½

		Contra Cr	£	S	D
1893		Brought forward	112	3	8
Jany	11	5 cwts Linseed cake 5 Cotton 1 Bran	3	18	6
Feby	2	Fire Insurance	"	18	"
		Expenditure	£117	"	2
		Income	87	19	6½
		Excess of Expenditure over Income	29	"	7½
		Feby 23rd 1893			
		Arch Stewart			

1893		Stock Dr to Cash	£	S	D
March	24	Sold 1 Fat Hiefer	18	"	"
April	5	1½ Quarters corn at 18/ per Qu	1	7	"
May	10	Sold 2 Stots	29	10	"
"	29	To wintering Horse	5	"	"
June	2	Sold 24 Weather Hogs at 26/ each	31	4	"
"	6	" 3 Quarters 2 bu at 14/ per Quarter	2	5	6
"	"	" 3 do at 17/	2	11	"
"	"	" 3 do at 18/	2	14	"
"	"	" 2 do 5 bu at 18/	2	7	3
Sept	11	" Wool	12	14	"
"	"	" 1 Ewe 42 lbs at 6d per lb	1	1	"
"	"	" 1 do 36 " " 6d " "	"	18	"
"	"	" 1 do 42 " " 6d " "	1	1	"
"	"	" 1 Ram	1	7	1
Oct	12	Sold 20 Ewes at 18/9 each	18	15	"
"	"	" 31 Weather Lambs at 9/9	15	2	3
"	"	2½ Quarters old Corn at 19/ per Qu	2	7	6
Dec	5	Sold 6 Ewes at 12/3 each	3	13	6
			£151	18	1

1893		Contra Cr	£	S	D
		Difficencie Brought forward from			
		last year	29	"	7½
March	21	Carriage of Meal & Cake	"	4	"
April	11	Carriage of Bran & Clover Seed	"	1	"
"	"	Seeds & Plants for Garden	"	2	6
"	"	C Mitchell & Son	1	4	9
"	"	Bought 1 Quarter Grass seed James Innes	1	12	"
"	"	" Seed Potatoes	"	11	8
May	29	David Moir Blacksmith a/c	1	9	½
"	"	Annie Low wages	7	"	"
"	"	Earl Dalhousie Rent	7	10	"
"	"	Engaging Servants	"	2	"
"	31	Joseph Dunbar wintering 32 Hogs	12	13	4
June	1	John Hood Coal & Cartage	1	7	"
"	3	D Stewart wintering 24 Hogs Carriage &			
		Commission	26	2	8
"	6	D & J Caithness seed Corn & flour	3	7	6
"	"	To wintering 87 old Sheep	10	6	8½
"	"	Bought one pig	"	19	"
"	"	To Niter for Horses	"	2	6
"	26	To P Peebles oat Meal	5	18	"
"	"	To W Napier wire Turnip scull	"	3	6
"	"	To R Middleton 1 day at peats	"	2	"
			109	19	9½

1893	Stock Dr to Cash	£	S	D
	Brought Forward	151	18	1

	£151	18	1

1893		Contra Cr	£	S	D
		Brought forward	109	19	9½
Sept	27	Nails 1/ & paint 6d	"	1	6
"	"	Coir Rope	"	2	2
Oct	12	Expenses & Commission at Brechin Sail	"	15	10
"	26	Brechin Agricultural Co Cake & Manure	2	18	"
"	"	Henderson & Son Clover Seeds & Plough line	1	8	10
"	"	John Hood Lime & Cartage	2	5	8
"	"	G Michie Summering 62 lambs 7 weeks a[t] 2½d per week	4	10	5
"	30	J Pirrie 3 Stirks 23 weeks at 1/6 each per week	5	3	6
Nov	24	David Moir Blacksmith	"	16	9
"	28	Ann Low wages	7	"	"
"	"	Archie Stewart herd	3	"	"
"	"	Earl of Dalhousie Rent	7	10	"
"	"	Mr Peebles Miller for Meal	3	12	"
"	30	David Mitchell wright	1	16	1
"	"	Engaging Servants	"	2	"
Dec	2	Poor & School Rates	"	7	7
"	"	Edward Duke Keep of 2 Rams	"	4	"
"	"	Bought 2 Rams at 30/ each	3	"	"
"	"	C Mitchell & Son Brechin	1	4	9
"	"	W Napier Ironmonger Brechin	"	9	11
			156	8	9½

1894		Stock Dr to Cash	£	S	D
Feby	24	Brought Forward	151	18	1
			151	18	1

1893		Contra Cr	£	S	D
		Brought forward	156	8	9½
Dec	26	County Rates	"	4	10
"	"	Geo Michie 2 Cows serving at 4/ each	"	8	"
"	"	G Michie Summer Dip for Old Sheep			
		& Lambs	"	8	"
1894					
Jany	6	Bought 2 lots of Blown Trees at Tarfside	"	8	"
"	"	2 Bottles Machien Oil		1	4
Feby	2	Fire Insurance	"	18	"
		Expenditure	£158	16	11½
		Income	£151	18	1
		Excess of Expenditure over Income	£6	18	10½

Feby 24th 1894

Archd Stewart

1894		Stock Dr to Cash	£	S	D
April	27	Sold 3 Bullocks	42	10	"
May	19	5 Quarters Corn at 17/6 per Qu	4	7	6
June	4	Wintering of Horse	5	"	"
Sept	11	586 lbs wool at 5¼d per lb	12	16	4
		1 Ewe	1	2	7
		1 do	1	3	8
		1 do	1	2	10
		1 do	1	6	8
Oct	6	20 Cast Ewes at 19/3 each	19	5	"
"	"	5 Skins	"	4	3
"	25	31 Lambs at 13/3 each less 5/			
		Commission	20	5	9
"	"	2 Quarters Corn at 18/ per Quarter	1	16	"
"	"	2 do do " "	1	16	"
Dec	18	to 12 Stones of Pork at 5/ per Stone	3	"	"
1895					
Jany	15	11 Weathers at £1.2/ each	12	2	"
		Income	£127	18	7
		Expenditure deducted	121	6	7½
		Funds on hand at date	6	11	11½
		Feburary 23rd 1895			
		Archd Stewart			

1894		Contra Cr	£	S	D
		Difficencie brought forward from last year	6	18	10½
April	3	Bought 2 young Pigs at 20/ each	2	"	"
"	14	1 Calf	1	"	"
"	"	2 Bottles Machien Oil	"	1	4
May	3	to 31 Hogs wintering per D Stewart	14	7	2
"	11	John Milne & Co Manure	3	18	"
"	28	Annie Low wages	7	10	"
"	"	Archie Stewart boy	1	"	"
"	"	1 Quarter Grass seed James Innes	1	10	"
"	"	Engaging Servants	"	2	"
"	"	Earl of Dalhousie six month rent	7	10	"
"	"	George Duncan 3 days work at Garden	"	4	6
"	"	D & J Caithness meal	3	12	"
"	"	David Moir Blacksmith	4	5	7
June	8	David Mitchell Joiner	"	16	7
"	16	Andrew Christison 5 days at peats	"	12	6
"	"	David Caithness wintering Ewes	11	"	"
"	19	C Mitchell & Son	1	4	9
July	17	D Ritchie Sheep dip	"	6	6
Augt	1	Annie Low wages to date	2	10	"
"	16	Robert Middleton 4 days at 2/ per day	"	8	"
			£70	17	9½

1894		Contra Cr	£	S	D
		Brought Forward	70	17	9½
Augt	16	Jane Eggo 15 days	"	15	"
"	22	Thatch for Houses	2	9	7
Sept	18	George Duncan 2 days at Harvest	"	3	6
"	"	Nails & paint	"	2	"
Oct	25	Expenses at Sail	"	10	"
"	"	to wintering 2 Tups & cover per Alex Stewart	1	10	"
"	"	Henderson & Son Seedsmen	1	7	8½
"	"	Brechin Agricultural Co Cake & Bran	5	2	"
Nov	2	to Summering 3 Stirk John Pirrie	5	2	"
"	"	John Milne & Co Turnip Manure	3	3	"
"	"	John Milne & Sons Turnip Seed	"	7	9
"	28	Griffeth Stewart wages	3	"	"
"	"	Bella Hunter do	3	10	"
"	"	David Caithness 50 posts for March fens	"	14	7
"	"	Earl Dalhousie Six month rent	7	10	"
"	30	David Moir Blacksmith	2	9	3
Dec	1	Poor & School Rates	"	6	5½
"	8	Stewart Porter for New Can to K Lum	"	5	"
			£109	7	7½

1894		Contra Cr	£	S	D
		Brought Forward	109	7	7½
Dec	5	To 76 Lambs Summering & Dipping			
		J Innes	7	4	4
"	17	To W Napier Ironmonger	"	18	2
"	"	2 Bolls meal Peter Peebles	1	13	"
"	18	County Rates	"	5	6
1895					
Feby	2	Fire Insurance	1	4	"
"	"	Dr Campbell 3 Cows serving at 5/ each		15	
			£121	6	7½

1895		Income	£	S	D
Feby	23	Income Brought Forward	6	11	11½
April	8	Sold 6½ Quarters Corn at 16/6 per Qu	5	7	3
May	2	" 2 Bullocks at £14.17.6 each	29	15	"
"	"	10 Bushels of Grass seed at 2/3 per bu	1	2	6
"	28	Wintering of Horse Lord Hindlip	5	"	"
Sept	9	520 lbs of wool at 5¼	11	5	"
Oct	7	to Ballance of Millers a/c for corn	4	16	8
"	10	26 weather Lambs at 15/6 each less commission 5/	19	18	"
"	"	20 Ewes & 10 Gimmers at 19/6 each	29	5	"
"	"	1 Ram	1	"	"
1896					
Jany	17	Sold 1 Fat Cow	18	10	"
		Income	£132	11	4½
		Expenditure deducted	116	11	5
		Funds on hand at date	£15	19	11½

Feby 4 1896 Archd Stewart

1895		Contra Cr	£	S	D
April	20	to David Stewart wintering 43 Hogs	18	14	5½
"	"	Plants to Garden		3	6
May	2	Expenses & Commission at Sail	"	15	"
"	"	C Mitchell & Son	1	13	"
"	"	Bought 2 Pigs at 18/ each	1	16	"
May	24	David Moir Blacksmith	1	6	6
"	"	David Caithness wintering Ewes	12	"	"
"	28	Bella Hunter six month wages	5	"	"
"	"	Engaging Servants	"	2	"
"	"	1 lb Turnip seed	"	"	8
"	"	Niter for Horses	"	2	6
"	"	Earl of Dalhousie six months rent	7	10	"
"	"	Bought 1 lot of Trees at Tarfside		3	6
"	"	2 Quarters seed corn at 18/ per Qu	1	16	"
June	5	R Middleton 3 days at peats		7	6
July	16	Brechin Agricultural Co Cake & Manure	4	6	"
"	"	D Ritchie Sheep Dip	"	10	6
"	"	Mr Bisset	"	1	6
Augt	12	P Brown repairs to Mill	2	14	9
"	23	Henderson & son seedsmen	1	3	3
June	18	Sheep Dip	"	7	8
"	"	A Christison carrier	"	3	7
			£60	17	10½

1895		Contra Cr	£	S	D
			60	17	10½
June	25	A Christison washin Tub	"	6	3
"	"	Carriage of Cake	"	"	9
Augt	2	Dip for Lambs	"	3	6
"	19	Carriage of Cake & Wool bags	"	1	"
"	23	one Bag of Dust	"	2	3
Sept	17	Mrs Duncan Harvest	1	13	6
Oct	10	1 Bag sement	"	6	"
"	"	Expenses at Brechin Sale	"	5	"
"	"	Turnip Manure	3	3	"
"	"	2 Rams 1 £1.7.6 one £2.10.	3	17	6
"	"	Expenses	"	1	"
"	"	Bought at whiggington Turnip	"	10	"
"	"	Barrow	"	1	"
"	"	Ladder	"	4	6
"	"	Stack Oats	3	"	"
"	24	J Hood Coal & cartage Manure	"	19	1
"	"	C Mitchell & Son	"	16	6
"	"	J Stewart 2 days at potatoes	"	4	"
		John Pirrie Summering 3 Stirks	5	5	9
		John Innes Summering Lambs	4	"	"
			£85	18	5½

		Contra Cr	£	S	D
		Brought Forward	85	18	5½
Oct	29	K Robertson 1 ton E Coal per Engine	1	6	10
Nov	27	David Moir Blacksmith	1	6	5
"	28	Earl Dalhousie Six Month rent	7	10	"
"	"	Thomson Saddler Edzell	"	5	6
"	"	Bella Hunter wages	5	"	"
"	"	Griffeth Stewart wages	3	"	"
"	"	Dr Campbell 1 cow served	"	5	"
"	"	Engaging Servant	"	2	"
"	"	Expenses of Rams coming home	"	2	"
Dec	9	William Naper Ironmonger Brechin	"	6	6½
"	12	David Mitchell Carpenter W Migvie	"	18	1
"	"	Brechin Agricultural Co Linseed cake	1	9	"
"	17	County Rates	"	5	2
"	20	J Milne & Sons Turnip seed	"	6	2
"	25	Poor & School Rates	"	7	"
1896					
Jany	6	Dr Ironside	1	12	"
"	"	Medicine for Cow & Horse	"	3	6
Feby	4	Brechin Agricultural Co Cotton Linseed &			
		Bran	5	3	9
"	"	Fire Insurance	1	4	"
			£116	11	5

1896		Income	£	S	D
		Brought Forward	15	19	11½
April	27	Sold 3 two year olds	40	10	"
June	16	To Ballance of Millers a/c for Oats	7	15	"
"	"	wool to Alexr Stewart	1		
"	"	wintering of horse Lord Hindlip	5		
		wool to Mr Scott	15	12	"
Oct	15	To 25 Ewes at 20/9d 10 Gimmers at 17/6)			
"	"	8 weathers at 22/6 less commission 10/9)	43	3	"
"	"	20 Lambs at 10/9d less commission 2/6	10	12	6
			£139	12	5½

£139 12 5½

1896		Contra Cr	£	S	D
April	16	to 100 Larch poasts	"	17	"
"	17	David Stewart wintering 40 Hogs	17	10	"
May	28	David Caithness wintering Ewes	11	18	4
"	"	Earl of Dalhousie Six Months rent	7	10	"
"	"	Robina Middleton Six months wages	7	10	"
"	"	Engaging Servants	"	2	"
"	"	David Moir Blacksmith	1	13	9
"	"	David Mitchell Joiner		6	6
"	"	Archd Stewart wages 28 weeks at 10/ per week	14	"	"
"	"	Bought 2 young pigs at 17/ each	1	14	"
"	"	Carriage of manure to Milden	"	8	"
"	29	William Glass for Moles	"	2	6
July	14	David Ritchie Sheep Dip	"	10	6
"	"	W Black & Son 3 Flakes & 1 harrow &c	"	15	"
"	"	C Mitchell & Son	1	13	"
"	"	Ferguson & Hood 1 Stable Broom		3	3
"	"	John Hood Coal & Cartage	1	8	5
"	"	Alexander Stewart wintering 2 Tups	"	17	"
Oct	15	Henderson & Son Seedsmen	1	4	5½
"	"	Bought at Sail of Turnabrain Hay &c	3	3	7
"	"	Expenses & Keep of Sheep at Sale		6	6
			£73	13	9½

1896	Income	£	S	D
	Brought Forward	139	12	5½

		£139	12	5½

1896		Contra Cr	£	S	D
		Brought Forward	73	13	9½
Oct	17	Mrs Duncan 2 days at Potatoes		2	
"	"	Arch Stewart Summering 3 S[t]irks	5	5	"
Nov	1	R Middleton 3 days Thatching Houses	"	6	"
"	21	Expenses of Rams coming home	"	5	"
"	26	James Lyall Blacksmith a/c	"	12	2
"	28	Jane Henderson Six months wages	7	10	"
"	"	Griffeth Stewart do do	3	10	"
"	"	Earl of Dalhousie six Month rent	7	10	"
"	"	David Caithness Miller for Meal &c	3	"	"
"	"	Mrs Duncan 4 days lifting Turnip	"	4	"
"	30	David Mitchell Joiner	"	16	1
Dec	7	Ferguson & Hood Sheep Dip &c	1	"	2
"	"	Poor & School Rates	"	7	"
"	"	County do		5	1½
"	17	John Milne & Co Turnip Manure	4	16	"
"	"	to Mother for Household expenses	2	"	"
"	"	1 Bottle of Machi[n]e Oil & Nails		2	6
"	"	David McLaren Straw for Thatch	3	13	9
"	"	2 Cows serving at 4/ each	"	8	"
			116	6	7

1897	Income	£	S	D
	Brought Forward	139	12	5½

		£	S	D
	Income	£139	12	5½
	Expenditure Deducted	120	16	2
	Funds on hand at date	£18	16	3½

Feby 10th 1897 Archd Stewart

1897		Contra Cr	£	S	D
		Brought Forward	116	6	7
Janry	5	Henry Meikeljohn Soap & oil for Deip		4	1
Feby	2	Expenses of Ewes to Glasterlaw	"	9	"
"	"	Alexr Stewart 2 Rams	2	12	6
"	"	Fire Insurance	1	4	"
			£120	16	2
			£120	16	2

1897		Income	£	S	D
		Brought Forward	18	16	3½
March	3	Sold 3 Fat Cattle	45	10	"
May	6	Sold 1 Stot £13.12.6 les commisson 3/6	13	9	
June	8	Sold 10 Eewe Hoggs at £1 each less			
		comm 2/6	9	17	6
Augt	10	Sold wool 591 lbs at 5⅛ 1 skin	12	13	11
Oct	14	Sold 24 Ewes at 24/3 each	29	2	"
"	"	" 30 BF Lambs at 13/ each	19	10	"
			£148	18	8½

		Contra Cr	£	S	D
1897		Brought Forward	72	2	6
"	"	Robert Campbell wintering Ewes	9	12	6
Augt	4	Alex Stewart wintering Rams &c		18	11
Oct	14	Commission on Sheep Farmers Mart	"	12	"
"	16	House & Harvest expenses &c paid by			
		William	4	11	5
"	18	W Keith Pigs Trough & Tar	"	8	6
"	26	Henderson & sons Seedsmen Brechin	1	7	10
"	"	C Mitchell & son		16	6
"	"	Brechin Agricultural & Turnip manure &			
		cake	4	8	6
"	"	Archd Stewart Summering 3 Stirks	5	3	6
"	27	Bought 1 Horse East Mains Balfour	11	2	3
"	"	Robert Middleton 1 day reparing Houses	"	2	"
Nov	22	George Duncan driving Rams	"	5	"
"	28	Susan Grant wages	8	"	"
"	"	James Mearns do	3	"	"
"	29	Earl of Dalhousie Six Months rent	7	10	"
"	"	John Shepherd Blacksmith	1	18	6
Dec	2	Poor & School Rates		2	9½
"	3	David Mitchell Carpenter	2	10	2
"	4	Mrs Duncan 6 days work at Turnips	"	6	"
"	11	County Rates	"	2	"
			£135	0	10½

1898		Income	£	S	D
Feby	2	Brought Forward	148	18	8½
		Expenditure deducted	136	10	4½
To Ballance of Funds on hand at date			£12	8	4

Feby 2 1898

Archd Stewart

		Contra Cr	£	S	D
1897		Brought Forward	135	0	10½
Dec	11	Stewart Porter 1 Bag of Cement	"	5	6
1898					
Feby	2	Fire Insurance	1	4	"
			£136	10	4½

1898		Income	£	S	D
		Brought Forward	12	8	4
March	24	Sold 2 Stots at £13 each	26	"	"
"	28	1 Fat Hiefer	17	15	"
May	16	1 Cow	9	10	"
"	"	9 BF Weathers	11	"	"
Augt	12	Wool	12	"	"
Oct	13	25 Ewes at 25/6 each	31	17	6
"	"	6 Gimmers 25/6 Do	7	13	"
"	"	30 Lambs 11/ Do	16	10	"
"	23	1 Cow £11.5/ less commission 2/9	11	2	3
"	31	1 Foal	9	"	"

1898		Contra Cr	£	S	D
April	15	Robert Duncan winter pasture for Ewes	8	"	"
"	"	Household expenses	1	"	"
"	22	George Stewart wintering Hogs	15	16	9
"	"	Plants & seeds for Garden	"	4	"
"	25	Bought 2 young Pigs at 17/6 each	1	15	"
May	28	John Shepherd Blacksmith	2	17	5
"	"	Susan Grant wages	8	"	"
"	"	Archd Stewart "	13	"	"
"	"	Earl of Dalhousie Rent	7	10	"
"	"	Hugh Campbell & Son 11 lbs Tea	1	2	"
"	31	C Mitchell & Son	1	4	9
"	"	William Napier Corn Bruiser	2	10	"
June	8	Bought 1 Cow in Calf	14	"	"
"	10	John Fraser Saddler	"	2	4
"	11	D Middleton 4 days at Peats	"	10	"
"	"	Engaging Servants	"	2	"
"	12	to 4 days keep of Cow at Edzell	"	8	"
"	"	Mrs Duncan 14 days at 1/ per day	"	14	"
"	16	David Mitchell Joiner	"	7	7
"	20	Miss Duke Tarfside	"	5	"
"	"	William Glass Moles	"	3	"
			£79	11	10

1898		Contra Cr	£	S	D
		Brought Forward	79	11	10
July	12	James Marshall Calf Meal	"	12	"
"	"	2 Bags Coal	"	4	6
Sept	22	Mrs Duncan Harvest work	"	17	"
"	"	J C Robertson Groceries	3	7	4
Oct	4	Mrs Gall Nurse	2	"	"
"	"	James Stewart 20½ lbs dripping	"	7	"
"	13	Expenses & Commission at Sale	1	2	"
"	"	Wm Reith 2 cwts Cake	"	19	"
"	"	Stewart Porter making grave	"	8	"
"	22	Robert Duncan Summering mare & foal	4	"	"
"	"	Archd Stewart Summering 3 Stirks	4	14	6
"	23	Farmers Mart 3 Rams at 45/	6	15	"
"	"	J C Robertson Merchant Edzell	"	7	6½
"	"	Henderson & sons Seedsmen	"	17	10
"	"	C Mitchell & son	2	8	9
"	"	Wm Napier Ironmonger	2	7	"
"	"	W & D Alexander Stationer	"	10	6
"	"	Brechin Agricultural Co Cake & Tracle	6	3	4
"	"	Ferguson & Hood	"	5	7
"	"	Wm Smith Flesher	"	11	4
			£118	10	0½

		Contra Cr	£	S	D
		Brought Forward	118	10	0½
Oct	23	Expenses at Sale with Cow	"	8	"
"	29	John Milne & Co Turnip Manure	3	7	"
Nov	8	Alexr Stewart Ballance of a/c for Rams	2	12	"
"	12	David Mitchell Joiner for coffin	3	5	"
"	"	David Mitchell Joiner for work &c	1	9	2
"	"	James Gibson wages	5	"	"
"	"	Earl of Dalhousie Rent	7	10	"
"	"	John Shepherd Blacksmith	1	1	4
"	29	Poor & School Rates	"	2	9½
"	"	Engaging Servant	"	2	"

Glossary

Arc	(ark) a chest for storing grain, etc.
Bran	husks of grain separated from meal after grinding
Cast ewe	a reject ewe, usually sold off as of no further service
Cess	a local tax
Coir rope	rope made of coir (coconut fibre) yarn
Curry comb	a comb for rubbing down or dressing a horse
Dust	(dist) particles of meal and husks produced in grinding
Fan	a set of fanners, a winnowing machine
Ferow	(farrow) of an animal, having missed a pregnancy
Flake	a portable framework of slats, for sheep working
Gimmer	a year-old ewe, a ewe between its first and second shearing
Heifer	a young cow
Hog(g)	a yearling sheep, a young sheep from the time it is weaned till it is shorn of its first fleece
Howing	hoeing (of turnips)
Lum	chimney
Merk	a unit of currency, two-thirds of a pound
Quay	(quey) a heifer (see above)
Sowens	flummery, a dish of oat husks and fine meal steeped in water
Stirk	a young bullock
Stot	a young castrated ox or bullock
Tare	(usually plural), peas etc. used as fodder
Tup	a ram
We(a)ther	a castrated ram

95